OVERVIEW OF OIL PRODUCTION:

DEMULSIFICATION OF OIL/WATER USING PLANT AS ALTERNATIVE

BY

EDWIN FRANCE

TABLE OF CONTENT

INTRODUCTION .... - 3 -

   Heavy oil, also known as Heavy Oil Properties .... - 3 -

CHAPTER ONE .... - 7 -

   Petroleum: A Precious Natural Resource.................................... - 7 -

CHAPTER TWO... - 13 -

   Crude oil reservoir... - 13 -

CHAPTER THREE .... - 19 -

   Emulsification.. - 19 -

CHAPTER FOUR .... - 25 -

   Demulsification .... - 25 -

CHAPTER FIVE.. - 29 -

   Demulsification using plant as alternative................................ - 29 -

CHAPTER SIX... - 33 -

   Merits of Plant-Based Demulsification: .... - 33 -

   The Demerits of Petroleum:... - 34 -

   Overview of the Environmental and Social Challenges .... - 34 -

CHAPTER SEVEN .... - 39 -

   CONCLUSION... - 39 -

INTRODUCTION

Heavy oil, also known as Heavy Oil Properties

Extra heavy crude oil or bitumen, is a type of crude oil that has a higher viscosity and density compared to conventional crude oil. Its properties and recovery methods differ significantly, requiring specialized techniques for extraction, recovery, and transportation.

Viscosity: Heavy oil has high viscosity, making it thick and resistant to flow. This characteristic makes it challenging to extract and transport using conventional methods.

Density: Heavy oil is denser than conventional crude oil, containing a higher proportion of heavier hydrocarbons.

Sulfur Content: Heavy oil often contains higher sulfur content, requiring additional processing to meet environmental standards.

Reservoir Characteristics: Heavy oil is usually found in unconventional reservoirs such as oil sands or reservoirs with low permeability.

Recovery Methods:

Steam Assisted Gravity Drainage (SAGD): This method involves injecting steam into the reservoir to reduce the viscosity of heavy oil, allowing it to flow more easily and be extracted through production wells.

Cold Heavy Oil Production with Sand (CHOPS): In this method, the natural energy from the reservoir is utilized to produce heavy oil. Sand is mixed with the oil, improving its flow properties.

In-Situ Combustion: Oxygen or air is injected into the reservoir, initiating combustion, which reduces the viscosity of heavy oil and helps in its recovery.

Solvent-Assisted Methods: Solvents like propane or butane are injected into the reservoir to dissolve heavy oil, making it easier to extract.

Transportation Methods:

Pipeline Transportation: Heavy oil can be transported through pipelines, although the pipelines need to be specially designed to handle its high viscosity and density.

Rail Transportation: Heavy oil can be transported in specially designed rail cars, providing flexibility in reaching destinations where pipelines are not available.

Marine Transportation: Heavy oil can be transported via tanker ships, but it requires careful handling due to its thick consistency, which can cause challenges during loading and unloading.

Upgrading and Refining: Before transportation, heavy oil is often upgraded or refined to convert it into a lighter, more transportable product, such as synthetic crude oil.

It's important to note that the choice of recovery and transportation methods depends on the specific characteristics of the heavy oil reservoir and the economic feasibility of the techniques involved.

CHAPTER ONE

Petroleum: A Precious Natural Resource

Petroleum, commonly known as crude oil, is a naturally occurring, flammable liquid found beneath the Earth's surface. It is a vital natural resource that has shaped modern industrial society in profound ways. Comprising hydrocarbons, petroleum is a complex mixture of organic compounds that serves as the raw material for various energy products and industrial applications.

Formation and Composition:

Petroleum forms over millions of years from the remains of ancient marine plants and animals that settled on the ocean floor. The intense heat and pressure within the Earth transform these organic materials into crude oil. The composition of petroleum varies, containing different proportions of carbon, hydrogen, sulfur, nitrogen, and trace elements.

Extraction:

The extraction of petroleum involves drilling deep wells into underground reservoirs. Once a reservoir is located, drilling rigs extract the crude oil, which is then transported to the surface. Advanced technologies, such as seismic imaging and directional drilling, enhance the efficiency and accuracy of extraction processes.

Refining:

Crude oil is a complex mixture of hydrocarbons with varying molecular weights. Refining, a crucial process in the petroleum industry, separates these hydrocarbons into different fractions based on their boiling points. Distillation, cracking, reforming, and other refining techniques are employed to produce products such as gasoline, diesel, jet fuel, lubricants, and petrochemicals.

End Products:

Transportation Fuels: Gasoline, diesel, and aviation fuels derived from petroleum power cars, trucks, ships, and airplanes, facilitating global transportation.

Petrochemicals: Petroleum is a primary source of raw materials for the production of plastics, synthetic rubber, solvents, fertilizers, and a wide range of chemicals essential for various industries.

Electricity Generation: In some power plants, petroleum is burned to generate electricity, especially in regions where natural gas and coal are less accessible.

Heating: Petroleum-based products, such as heating oil, are used for residential and commercial heating purposes in colder climates.

Environmental Impact:

While petroleum products are invaluable, their extraction, transportation, and combustion have environmental consequences. Oil spills, air pollution, and greenhouse gas emissions contribute to environmental degradation and climate change. Efforts to develop cleaner technologies, reduce emissions, and transition to renewable energy sources aim to mitigate these impacts.

Global Importance:

Petroleum is a global commodity, with major oil-producing countries including Saudi Arabia, the United States, Russia, and China. The petroleum industry significantly influences geopolitics, economics, and international relations, often shaping global energy policies and alliances.

Future Outlook:

As the world transitions towards sustainable energy sources, petroleum remains a critical component of the energy mix. Research and development efforts focus on improving extraction techniques, enhancing refining processes, and developing alternative energy solutions to meet the world's growing energy demands while minimizing environmental impact.

In summary, petroleum stands as a cornerstone of modern industrial civilization, providing energy, fueling economic growth, and driving technological innovation. As society advances, the responsible and sustainable management of this precious resource is paramount for the well-being of

current and future generations.

CHAPTER TWO

Crude oil reservoir

A crude oil reservoir is a subsurface pool of hydrocarbons contained in porous or fractured rock formations. These reservoirs are crucial in the petroleum industry as they serve as the source of crude oil, which is subsequently extracted, refined, and processed into various products. Here are the key aspects of a crude oil reservoir:

Formation:

Crude oil reservoirs are formed over millions of years through the accumulation of organic materials, primarily the remains of ancient marine plants and animals. These organic materials settle on the seabed and are gradually buried under layers of sediment. The heat and pressure from the Earth's crust transform these organic materials into hydrocarbons, creating crude oil.

Reservoir Characteristics:

Porous Rock: The reservoir rock is typically porous, allowing the accumulation of crude oil within its interconnected pores. Common porous rocks include sandstone and limestone.

Permeability: Permeability refers to the ability of the rock to transmit fluids. Reservoir rocks need to have sufficient permeability to allow the movement of crude oil through the rock pores.

Cap Rock: A layer of impermeable rock, such as shale, forms a cap above the reservoir, preventing the upward migration of oil and gas. This cap rock traps the hydrocarbons within the reservoir.

Exploration and Identification:

Seismic Imaging: Advanced seismic techniques use sound waves to create detailed images of subsurface rock formations. Seismic surveys help geologists identify potential reservoir structures.

Exploratory Drilling: Exploratory wells are drilled to assess the presence of oil or gas. Core samples and well logs provide valuable information about the composition and characteristics of the reservoir.

Extraction:

Primary Recovery: Initially, the natural pressure within the reservoir forces the oil to the surface. This primary recovery method often utilizes the reservoir's natural energy to extract oil.

Secondary Recovery: Techniques like water flooding involve injecting water into the reservoir to maintain pressure and displace additional oil towards the production wells.

Enhanced Oil Recovery (EOR): Advanced methods, such as injecting steam, gases, or chemicals into the reservoir, are employed in EOR to extract more oil after primary and secondary recovery phases.

Reservoir Management:

Reservoir Modeling: Engineers create detailed models of the reservoir to understand its behavior, predict production rates, and optimize extraction techniques.

Pressure Maintenance: Techniques like gas injection or water flooding are used to maintain reservoir pressure, ensuring efficient oil recovery.

Environmental Considerations: Reservoir management also involves addressing environmental concerns, such as minimizing the risk of oil spills and groundwater contamination.

Crude oil reservoirs are invaluable assets, and their effective management requires a combination of geological understanding, engineering expertise, and environmental stewardship. The exploration, extraction, and sustainable use of these reservoirs are essential for meeting global energy demands while minimizing environmental impact.

CHAPTER THREE

Emulsification

Emulsification is a process in which two immiscible substances, such as oil and water, are mixed together to form a stable and homogeneous mixture called an emulsion. Emulsions are widely used in various industries, including food, cosmetics, pharmaceuticals, and petroleum, due to their ability to combine substances that do not normally mix. Here's a closer look at emulsification and its applications:

Process of Emulsification:

Emulsifying Agent: An emulsifying agent, also known as an emulsifier, is necessary to stabilize the emulsion. Emulsifiers have molecules with one hydrophilic (water-attracting) and one hydrophobic (oil-attracting) end. These molecules surround the droplets of one substance in the mixture, preventing them from coalescing.

Mixing: Emulsification involves vigorously mixing the immiscible substances along with the emulsifying agent. This mechanical action breaks down the larger droplets into smaller ones, dispersing them throughout the mixture.

Stabilization: The emulsifying agent stabilizes the emulsion by forming a protective layer around the droplets. This prevents the droplets from recombining and separating back into their original phases.

Types of Emulsions:

Oil-in-Water (O/W) Emulsion: In this type, oil droplets are dispersed within a water-based solution. Milk is a common example of an O/W emulsion, where tiny oil droplets are dispersed in water.

Water-in-Oil (W/O) Emulsion: In this type, water droplets are dispersed within an oil-based solution. Butter is a natural example of a W/O emulsion, where water droplets are dispersed in the fatty matrix.

Applications of Emulsification:

Food Industry: Emulsions are widely used in food preparation. Mayonnaise, salad dressings, and sauces are O/W emulsions. Margarine and certain types of chocolates are examples of W/O emulsions.

Cosmetics and Personal Care: Emulsions form the base for many cosmetic products. Creams, lotions, and moisturizers are O/W emulsions, while cold creams and ointments are W/O emulsions.

Pharmaceuticals: Emulsions are used in pharmaceutical formulations, enabling the combination of oil-based and water-based medications for oral or topical administration.

Petroleum Industry: Emulsification is used in enhanced oil recovery techniques, where chemicals are injected into reservoirs to modify the properties of crude oil, making it easier to extract.

Paints and Coatings: Emulsions are utilized in the formulation of paints and coatings, ensuring a smooth and uniform application on surfaces.

Detergents: Emulsifiers in detergents help in the removal of greasy stains by forming emulsions with oil and facilitating their removal from fabrics.

Emulsification plays a fundamental role in various industrial processes, allowing for the creation of stable mixtures that would otherwise separate into distinct layers due to the

immiscibility of their components.

CHAPTER FOUR

Demulsification

Demulsification is the process of separating emulsified substances, typically oil and water, into their individual components. Emulsions are stable mixtures of immiscible liquids, such as oil and water that are held together by an emulsifying agent. Demulsification involves breaking this emulsion, allowing the oil and water to separate from each other.

Methods of Demulsification:

Chemical Demulsifiers: Chemical demulsifiers are substances specifically designed to break emulsions. They work by disrupting the emulsifying agent's ability to hold the oil and water together. These demulsifiers can be added directly to the emulsion, causing the droplets to coalesce and separate.

Heating: Heating an emulsion reduces its viscosity, making it easier for the oil and water droplets to coalesce and separate. Heat destabilizes the emulsifying agents, allowing the natural tendency of oil and water to separate based on their densities.

Centrifugation: Centrifugation involves spinning the emulsion at high speeds in a centrifuge. The centrifugal force causes the denser phase (usually water) to move outward, allowing the separation of oil and water layers.

Electrocoagulation: This method uses an electrical field to destabilize the emulsion. Charged particles within the emulsion coalesce, leading to the separation of oil and water.

Ultrasonic Treatment: Ultrasonic waves can disrupt the stability of emulsions, causing the droplets to merge and separate. Ultrasonic demulsification is particularly useful for breaking stable and fine emulsions.

Applications of Demulsification:

Oil Industry: In the petroleum industry, crude oil often contains water and emulsified impurities. Demulsification is crucial for separating the oil and water phases, allowing the purification of crude oil before refining.

Wastewater Treatment: Demulsification is used to treat industrial wastewater containing oil-water emulsions. Separating the oil from water allows the safe disposal or reuse of treated water.

Food and Beverage Industry: Demulsification is employed in processes like dairy production to separate cream from milk, and in the production of salad dressings to prevent the separation of oil and water-based ingredients.

Chemical Processing: Demulsification is used in chemical processes where emulsions need to be broken for purification, product recovery, or efficient processing.

Environmental Cleanup: In the case of oil spills, demulsification agents are used to break down emulsified oil, facilitating its removal from water surfaces.

Demulsification techniques are essential for various industries, enabling the efficient separation of emulsified liquids and ensuring the quality and purity of the separated components.

CHAPTER FIVE

Demulsification using plant as alternative

Demulsification using plant-based methods involves utilizing natural compounds found in plants to break down emulsified substances, such as oil and water. Certain plants contain bioactive compounds that can act as demulsifiers, disrupting the stability of emulsions and promoting the separation of oil and water phases. Here are a few ways plant-based materials can be used for demulsification:

1. Plant Extracts:

Certain plant extracts, rich in organic compounds like saponins and flavonoids, have demulsifying properties. These extracts can be mixed with emulsified solutions to facilitate the separation of oil and water. For example, extracts from plants like soapnut (Sapindus mukorossi) have natural surfactant properties, making them effective demulsifiers.

2. Enzymes:

Enzymes derived from plants can also be used for demulsification. Enzymes are biological molecules that can break down complex substances. Some plant-based enzymes can disrupt the emulsifying agents, causing the emulsion to destabilize and separate. Protease enzymes found in certain plants are known for their emulsion-breaking capabilities.

3. Essential Oils:

Essential oils obtained from plants contain volatile compounds that can interfere with the stability of emulsions. Essential oils like citrus oils (e.g., orange, lemon) have been studied for their demulsifying properties. These oils can be added to emulsions to promote the separation of oil and water phases.

4. Polysaccharides:

Certain plant-derived polysaccharides, such as gum arabic from the Acacia tree, have natural emulsifying and demulsifying properties. These polysaccharides can be used to stabilize or destabilize emulsions based on the specific conditions and concentrations used.

5. Green Surfactants:

Researchers are exploring the use of green surfactants derived from plant sources for demulsification purposes. These surfactants are eco-friendly alternatives to synthetic chemicals and can be effective demulsifiers, especially in mild conditions.

CHAPTER SIX

Merits of Plant-Based Demulsification:

Environmentally Friendly: Plant-based demulsification methods are generally environmentally friendly and biodegradable, reducing the environmental impact of chemical processes.

Renewable Source: Plants are a renewable resource, making plant-based demulsification sustainable in the long term.

Safety: Plant-based demulsifiers are often safer to handle and use, especially in applications involving food, beverages, and cosmetics.

It's important to note that the effectiveness of plant-based demulsification methods can vary based on the specific emulsion composition and conditions. Research in this area continues to explore the potential of plant-derived compounds for efficient and sustainable demulsification processes.

The Demerits of Petroleum:

Overview of the Environmental and Social Challenges

Petroleum, a finite fossil fuel, has undeniably propelled the progress of human civilization. However, its widespread use comes at a significant cost, encompassing environmental degradation, geopolitical tensions, and socioeconomic disparities. Understanding the demerits of petroleum is crucial for fostering awareness, encouraging sustainable practices, and exploring alternative energy sources. In conclusion, while petroleum has been a cornerstone of progress, its demerits cannot be ignored. Addressing these challenges requires a concerted effort from governments, industries, and individuals alike. Embracing renewable energy sources, investing in research and development, promoting energy efficiency, and adopting responsible extraction practices are crucial steps toward mitigating the demerits of petroleum. By fostering awareness and taking proactive measures, we can pave the way for a more sustainable and balanced energy future.

1. Environmental Pollution:

One of the most significant demerits of petroleum lies in its environmental impact. The combustion of petroleum products releases greenhouse gases, contributing to climate change and global warming. Additionally, emissions from vehicles and industrial processes lead to air pollution, smog, and respiratory problems, impacting both human health and the natural environment.

2. Depletion of Natural Resources:

Petroleum is a finite resource, and the extraction process often involves invasive methods like drilling and fracking, leading to habitat destruction and biodiversity loss. As reserves deplete, the industry resorts to more challenging extraction techniques, intensifying environmental disturbances.

3. Water Pollution:

Oil spills, a common occurrence in petroleum extraction and transportation, pose a severe threat to aquatic ecosystems. These spills contaminate water bodies, harming marine life, disrupting ecosystems, and affecting coastal communities that rely on fishing and tourism. Cleaning up oil spills is an arduous and expensive process, often leaving long-lasting ecological scars.

4. Geopolitical Tensions and Energy Security:

Petroleum-rich regions often become focal points of geopolitical tensions. Control over oil reserves can lead to conflicts and power struggles among nations, impacting global stability. Moreover, dependence on petroleum imports creates vulnerabilities in a country's energy security, leaving economies susceptible to price fluctuations and supply disruptions.

5. Social Disparities and Economic Challenges:

While petroleum extraction can bring economic opportunities to regions, it also exacerbates social disparities. Local communities, especially in developing countries, often face displacement, loss of livelihoods, and adverse health effects due to petroleum-related activities. Additionally, fluctuations in oil prices can destabilize economies, impacting employment rates and government budgets.

6. Non-renewable Nature:

Petroleum is a non-renewable resource, meaning it cannot be replaced once depleted. As the global demand for energy continues to rise, the finite nature of petroleum poses a significant challenge. This necessitates a shift toward renewable and sustainable energy sources to meet future energy needs.

CHAPTER SEVEN

CONCLUSION

In the grand tapestry of human progress, few resources have played as pivotal a role as petroleum. From fueling the engines of our transportation systems to being the building blocks of countless everyday products, petroleum has been the lifeblood of modern civilization. However, as we stand on the cusp of a new era, it's essential to reflect on the past, acknowledge the present challenges, and envision a sustainable future.

Reflection on the Past:

Petroleum, in its various forms, has powered the world for well over a century. It has driven economies, shaped geopolitics, and transformed societies. The oil industry has pioneered technological advancements, from drilling techniques that reach unimaginable depths to refining processes that yield cleaner fuels. Our dependency on petroleum has undeniably fueled global progress, albeit at a cost.

Acknowledging Present Challenges:

As we move forward, it's crucial to confront the environmental, social, and economic challenges associated with petroleum. Climate change, air pollution, and habitat destruction are among the environmental concerns, urging us to transition towards renewable and sustainable energy sources. The social impact of oil extraction on local communities and indigenous populations cannot be ignored. Additionally, the volatile nature of the oil market poses economic challenges for both producing and consuming nations.

Envisioning a Sustainable Future:

The future of petroleum lies in our ability to innovate, diversify, and embrace sustainable practices. Investments in renewable energy sources, such as solar, wind, and hydroelectric power, are essential for reducing our carbon footprint. Research and development in energy storage technologies can pave the way for efficient utilization of renewable energy. Moreover, promoting energy efficiency in all sectors of society can significantly reduce overall energy consumption.

Furthermore, the oil industry itself can contribute to a sustainable future through responsible extraction practices, minimizing environmental impact, and engaging with local communities to ensure fair and equitable benefits. Embracing circular economy principles, where products and materials are reused, recycled, and repurposed, can reduce the demand for virgin petroleum-based products.

In conclusion, the journey with petroleum has been remarkable, but it's imperative to evolve. Our ability to strike a balance between meeting energy demands, environmental stewardship, and social responsibility will define our path forward. By embracing innovation, international collaboration, and a collective commitment to sustainability, we can navigate the future of petroleum in a way that ensures a prosperous and harmonious coexistence with our planet.

As stewards of this precious resource, it is our responsibility to shape a future where petroleum, along with renewable energy sources, contributes to a sustainable, equitable, and

thriving world for generations to come